THE BUTTERFLIES IN CHRISTCHURCH PARK

Past, present and future

(Photo: Liz Cutting)

Richard Stewart

Photos by Liz Cutting

First Published in 2016 by
The Friends of Christchurch Park, Ipswich

www.focp.org.uk

text and photos © 2016 Richard Stewart
photos © 2016 Liz Cutting
photos © 2016 Peter Maddison
photos © 2016 Anne-Marie Stewart

All rights reserved. No part of this book may be reprinted or reproduced or utilised in any form or by any electronic, mechanical or other means, now known or hereafter invented, including photocopying and recording, or in any information storage or retrieval system, without permission in writing from the author.

A catalogue record of this book is available from the British Library

Stewart, Richard
The Butterflies in Christchurch Park

Editing and layout by Martin Sanford

Printed by Tuddenham Press Ltd.

ISBN 978-0-9564584-5-2

Other books in the series about Christchurch Park are:
Portrait of a Park
A year with the wildlife of Christchurch Park

Portrait of an Owl
My tales of Mabel and other owls

Portrait of the Birds
50 years of Birdlife in Christchurch Park, Ipswich

All three written and illustrated by Reg Snook, the 'Portrait of the Birds' in collaboration with Philip Murphy

Ipswich Arboretum - A History and Celebration
David Miller, with contemporary photos by Liz Cutting

For more details of these and other books published by
the Friends of Christchurch Park go to
www.focp.org.uk

A message from the Friends...

It gives me enormous pleasure to introduce Richard's *The Butterflies in Christchurch Park* as the Friends' latest publication. Richard first approached me with his proposal last March, within a week of my appointment as Chair at our AGM. I am so glad that he did and my Committee jumped at the opportunity to support another title for our wonderful Park. Richard could not have been better qualified for this project, for nearly a decade he was County Butterfly Recorder for Suffolk and he has published half a dozen previous books including *The Millennium Atlas of Suffolk Butterflies* in 2001 (Suffolk Naturalists' Society) and over 5,000 articles, poems and photos in local, national and international titles. He is a Naturalist in the truest sense who loves nothing more than to get out into the field to record his own observations.

I recall one afternoon last August when I was watering trees in the Arboretum, I bumped into Richard who was on his way to Ipswich Museum. We of course started talking about the book and just then a yellow butterfly flew past us. We looked at each other and I said "What was that?" Richard replied "That might have been a clouded yellow and I've never seen one in the Arboretum or Park." Before I could say anything more he was off, rushing down the path in hot pursuit of a possible clouded yellow...

In support of the launch of this book the Friends are delighted to be funding a substantial quantity of plants to improve the habitat for butterflies in the Park and Arboretum. These species have been selected by Richard and include common and alder buckthorn, gorse, broom, honeysuckle, aubretia, arabis, alyssum, thyme and marjoram.

We would like to congratulate Richard on an outstanding study and thank him for most generously donating his royalties to the Friends in order to support our Arboretum Armillary Sphere Sundial Restoration Appeal in memory of Dr John Blatchly MBE. We would also like to record our immense gratitude to Liz Cutting and Martin Sanford. Liz has yet again supplied us with a stunning set of photographs and Martin has assisted Richard throughout the entire process helping him to deliver a book from proposal to launch in less than eight months.

David Miller
Chairman, Friends of Christchurch Park
November 2016

CONTENTS

About the author and the photographer	6
Introduction	8
Butterfly habitats	12
Butterfly species recorded in the park	18
The future	48
Predators in the Park	56
Summary of butterfly species	60
Flight Periods	62
Creating a butterfly garden	64
Epilogue - butterfly haiku	74
Index	79

ABOUT THE AUTHOR

Richard Stewart was the Suffolk Butterfly Recorder from 1994-2002. During that time he organised county wide surveys of gardens and churchyards as well as the five year Millennium Survey. This produced his book 'The Millennium Atlas of Suffolk Butterflies'. His other butterfly book 'Butterfly Days' covered diary entries during this survey. He has also written a book about the life and writings of the renowned Suffolk botanist, Francis Simpson. Richard's work has appeared in 240 different publications. For many years he wrote regular monthly illustrated articles for the local 'Norfolk and Suffolk Life' and the national 'Cycling World'. Currently he writes regularly for eight different publications.

A passionate long term advocate of cycling, walking and public transport, wherever possible, Richard has covered the country to find all the resident butterfly species and three other rare migrants. Less environmentally friendly journeys have taken him to eight different Latin American countries, seeing amazing wildlife from jaguars in Brazil to large blue iridescent morpho butterflies in the rainforests. In his opinion such journeys are justified because they create many eco-tourism jobs for local workers and stop the destruction of habitats.

Richard has had three poetry collections published and his speciality is the haiku form. He has had over 1,000 haiku published and has won several national poetry prizes. His last book, 'Dark Sky Dancing' was a seasonal collection of haiku, illustrated by photos of his wife Anne-Marie's art quilts. They live just a few hundred yards from the park and their relatively tidy back garden is managed as a wildlife reserve. Over many years it has welcomed feeding waxwings, overwintering blackcaps, bathing sparrowhawks, families of hedgehogs, bank voles and woodmice as well as banded demoiselle damselflies and the spectacular hummingbird hawk moth. There is particular emphasis on a wide variety of nectar plants, to encourage many butterfly species into the garden.

SPECIES PHOTOS BY LIZ CUTTING

Liz took up photography in the mid-1990s. She is a keen naturalist and has been concentrating on nature photography for most of the last ten years. She also volunteers in the conservation sector, and in particular is very involved with dormouse conservation in Suffolk and Essex. A regular on the national and international photography exhibition circuit, Liz's images have been used by the RSPB, Suffolk Ornithology Group, Suffolk Naturalists' Society, Colchester Natural History Society, the Hawk and Owl Trust and the British Trust for Ornithology, and have appeared in local and national newspapers, in journals, on Christmas cards and in the *Bird Atlas 2007-11*.

(Richard Stewart)

Wide borders along the edge of the formal lawn leading up to Christchurch Mansion: one of the first butterfly habitats seen on entering the park from the Soane Street entrance.

INTRODUCTION

On July 3rd. 2016 I was standing on one of the higher areas in the park. This gave a panoramic view of the many thousands who were enjoying the annual Music Day. With this number of people, the many attractions present and the loud music from various stages, it may seem amazing that any wildlife could still be present. The reality is of course very different, as Reg Snook has so eloquently described in his regular and continuing notes about the park's natural history. Christchurch Park is approximately 82 acres in extent, a designated County Wildlife Site within a Conservation Area and registered with English Heritage for the special historic interest of its many different features.

A surprising number of butterfly species can be seen here, even though the centre of Ipswich is very close. Some pass through, some appear during migration, while others spend the whole of their amazing life cycle within the park, from egg to adult form. This book looks first at the many different butterfly habitats within the park then a longer section gives more details of any individual species recorded here, from those usually abundant to literally single sightings. A third section looks at the future, examining how extra nectar plants can be introduced. A short list of additional species that could subsequently be attracted is also included. Finally, at the suggestion of one member of the Friends' committee, I have explained how a good butterfly garden can be created.

I wish to thank David Miller, the current chairman of the Friends of Christchurch Park. He has been a valuable contact and support in the writing of this book. The rest of the committee also gave me their enthusiastic support when the idea was first presented to them. Both David Miller and Joe Underwood, the Education and Wildlife Ranger, have been instrumental in securing the necessary finance to provide extra nectar sources within the park. I also wish to thank Reg Snook for his pioneering work in publicising the park's wildlife through his recent books and regular wildlife log. Thanks also go to Steve Kemp for giving permission to reproduce the park map on which the habitats are numbered, and Sam Teague and Steve Leech for providing the necessary image to reproduce. Liz Cutting is already well known for her park photography, especially the contemporary images in David Miller's recent publication

'Ipswich Arboretum - A History And Celebration'. I contacted Liz well in advance of approval being given for this book and throughout the process she has been extremely helpful and supportive. Almost all of the butterflies featured were photographed by Liz. Peter Maddison, current chairman and newsletter editor of the Suffolk Branch of Butterfly Conservation, also contributed images and the rest are by my wife Anne-Marie and myself. My IT skills are marginally above Luddite so I am extremely grateful to Martin Sanford, who runs the Suffolk Biodiversity Information Service, at the Ipswich Museum. Without his patient help and advice this book would never have been published.

Richard Stewart, August 2016.

(Richard Stewart)

The sunlit garden at the back of Christchurch Mansion has a wealth of nectar sources, in particular these extensive lavender beds which attract summer butterflies, especially small, large and green-veined whites.

Numbered butterfly habitats within the park, see following text.

BUTTERFLY HABITATS

The accompanying map lists, by number, the main butterfly areas in Christchurch Park. However, butterflies can be seen almost anywhere, especially when good nectar sources such as dandelions, hawkweed and clover flower abundantly in grassed areas. Butterflies have literally been seen from the house hedge next to the Soane Street gate to the buddleia near the exit into Park Road.

1. This small garden under the large plane tree attracts early insects, including butterflies emerging from hibernation. Nectar sources present are daffodils, snowdrops, celandines, crocuses and red deadnettle. My first 2016 butterfly was seen here on 10th March - a small tortoiseshell perched on top of a snowdrop. The house hedge across the path has ivy, which attracts holly blues.

2. There are deep borders to the formal lawn in front of Christchurch Mansion. These have bluebells and red deadnettle, followed by longer grasses in summer. These grasses are used for egg-laying by many summer butterflies. The adjacent graveyard of Saint Margaret's church was the subject of a wildlife audit some years ago and I gave advice about improving its wildlife potential.

3. Small gardens with good nectar sources are close to the car park and also several clumps of ivy.

4. The garden at the back of Christchurch Mansion is a good sunspot. It has extensive summer beds of lavender, a few of the tall *Verbena bonariensis* and later flowering Michaelmas daisies. All are abundant in nectar.

5. The sloping bank next to the Reg Driver visitor centre has a hedge of native species; some of these, when in flower, offer nectar. Below, in July 2016, among the long grasses, were brambles and hawkweed, plus a few plants of yarrow and ragwort. This last mentioned plant can be dangerous to grazing animals but is rich in nectar. It is also used for egg-laying by the bright red day-flying cinnabar moth. The emerging black and yellow caterpillars can completely strip the foliage. This habitat was completely cut down to the ground in the first half of July 2016. This prevents the proper seeding of plants and the full life cycle of summer butterflies. Hopefully it will in future be cut at least one month later.

(Anne-Marie Stewart)

(Richard Stewart)

In the butterfly garden: the top picture was taken during a butterfly walk on 6th August 2016, when ten species were seen.

6. The extensive woodland reserve has many areas of deep shade that do not attract butterflies. However the main paths are wide and sunlit, good for basking butterflies and especially favoured by the speckled wood, which prefers dappled shade. Bordering the reserve are large clumps of bramble, some nettles, one short stretch of privet and both holly and ivy. The last two are the main plants used for egg-laying by holly blues.

7. This steeply sloping bank is behind what used to be the basketball court. There is one buddleia and among the longer grasses there is ragwort, bramble, hawkweed, clover and yarrow. This is a good location for the summer flying ringlet, which usually prefers damper habitats.

8. The butterfly garden is an open and sunny spot, offering a continuity of nectar from spring to autumn. Spring flowers include primulas and primroses then the early flowering *globosa* buddleia, plus scabious, lavender, erigeron, *Verbena bonariensis* and phlox. The purple *davidii* varieties of buddleia flower in summer and regular deadheading of florets can extend their flowering period. Some possible planting additions, particularly in spring, are discussed in the 'Future' section of this book. Other good nectar sources such as jack-by-the-hedge, hawkweed, common hemp nettle and clover have appeared over the years. Here there is an important series of continuing habitats. The banks are being colonised by lady's smock from the wet meadow and the tall laurels behind the garden have early flowers for butterflies needing nectar after emerging from hibernation, in particular peacocks. The extensive bramble bush has an abundance of flowers which are probably the main nectar source for summer insects. Butterflies also feed on the soft fermenting blackberries and the depth of cover is used for both roosting and hibernation. The nearby oak tree has a colony of the mainly arboreal purple hairstreaks.

9. This continuity of habitats also embraces the woodland reserve and longer grass areas, plus the mature holly trees along the path close to the butterfly garden and heading towards the tennis courts. Holly is used for egg-laying by one generation of holly blues and they can often be seen in this area.

10. This wet meadow was cut twice in 2015 and next spring there was an abundance of lady's smock. This is used for egg-laying by orange tip and green-veined white butterflies.

11. From here onwards large areas are composed of longer grasses. In their midst can be found hawkweed, yarrow and knapweed. The higher path, which is closest to the tennis courts, offers a good view downwards and this is the best place to see common blue butterflies. If you follow this higher path it passes bushes good for nectar and egg-laying: gorse, broom, bramble, laurel and holly.

12. The paths through longer grasses, leading up to the frog seat, are usually good for small coppers and meadow browns. The latter is probably the most common summer butterfly in the park. The oaks close to the frog seat have colonies of purple hairstreaks.

(Richard Stewart)

Habitat number 11 includes several nectar and food plants on the left of the path and longer grasses to the right, extending right up to the top north of the park.

13. To the left, the Upper Arboretum is a large and sunlit area but the more formal planting doesn't attract many species. On a July 2016 visit I noted a few oak trees and some holly but otherwise there was just a sprinkling of clover and hawkweed near the top of the long raised flower bed. However, this bed was my first choice for a butterfly garden in an Ipswich park during the 1990s, though replanted with more nectar sources. It was eventually located in Alexandra Park.

14. This northern corner has yarrow, hedge mustard and common hemp nettle. The main butterfly attraction comes from the luxuriant growth of nettles around an old log. Nettles are the plants used for egg-laying by the small tortoiseshell, peacock, comma, red admiral and occasionally the painted lady. There is also a buddleia in the nearby lane.

15. The nearby copse is too shaded in summer to attract many species. Earlier in spring it has celandines, crocuses and red deadnettle. With all the buildings and mature trees in the park, there are many good hibernation sites. These could also include the nearby stag beetle clump, one of several in the park.

16. This area is dominated by large drifts of longer grasses and several of the oaks have purple hairstreak colonies.

17. From the copse to the Park Road gate there are tall and sunlit bramble beds, extending several yards from the fence.

18. The Park Road entrance, at the park end, is dominated by tall and long established clumps of ivy. These attract many insects in early autumn as the nectar bearing umbels open, including hornets and red admirals.

19. At the road end of this entrance a tall buddleia attracts several species, including the migratory painted lady.

20. To the right of this entrance brambles continue and in the far corner is a recently planted hedge of native species which includes buckthorn, on which the brimstone lays its eggs. Many banks of longer grasses, mixed in with nectar plants, extend along the park border with Westerfield Road and for most of the boundary back to the Reg Driver visitor centre.

21. The recently created orchard includes a hedge of native species that can offer early nectar during their flowering period. Once the fruit trees become fully established their bounty may be exploited by butterflies such as the red admiral, comma and speckled wood. These use their long uncoiled proboscis to suck up the juices of fermented fruits, sometimes on the tree but usually on those fruits that have fallen. There have been recorded cases of both birds and butterflies becoming inebriated and finding it difficult to fly.

This habitat also extends the continuity described in comments about the butterfly garden, as it directly joins the woodland reserve.

(Richard Stewart)

Butterfly habitat number 17: extensive bramble beds when in sunshine offer one of the best summer nectar sources for a wide variety of insects.

BUTTERFLY SPECIES RECORDED IN THE PARK

Many of the park's butterfly species were briefly mentioned in the habitats section but now they are described in more detail. I have followed the normal order of species that will be found in almost all butterfly identification books and I have, with a very few exceptions, not used Latin names for individual species or indeed anything else. Similarly, I have used the term 'egg-laying' rather than the more obscure 'ovipositing' and referred to the other stages through to the adult butterfly by the words 'caterpillars' and pupae', both obviously plural. Some butterflies can have as many as four generations in a good breeding year but I have used the word 'broods' to cover this. Every species included has been seen at least once in the park during the years from 2003 to 2016. Records for this period were provided by Martin Sanford and have come from myself, my wife Anne-Marie, and all those listed here: David Basham, Matt Berry, Liz Cutting, Matthew Garnham, Jean Garrod, Steve Goddard, Stuart Gough, Julia Harrison, Ben Heather, Alan Johnson, Steve Lenane, Kevin Ling, Barbara Mathews, Martin Peers, Gary Plank, Helen Saunders, Mark Scutt, Reg Snook, Bill Stone, Joe Underwood, Julia Usher, Phillip Vaughan-Williams, Laura Whitfield and Stella Wolfe.

Small and Essex Skipper
Most species are described separately but these two 'golden' skippers are almost identical. Skippers are the most moth-like of our butterflies and in one reference book I have the author lists them among the moths. These are among the smallest of our Suffolk butterflies with a wingspan average 27-30 mm. Their second name is most appropriate as the flight pattern is a series of short hops or skips, often moving rapidly from one plant to the next and defying attempts to separate them by name. This is best done by examining the undersides of the antennae - reddish orange in the small skipper and deep black in the Essex. The angle at rest and worn specimens complicate identification, which is best done with close focus binoculars. Their development is different as the small skipper overwinters as a caterpillar and the other as an egg. Both have a single brood each year and are on the wing normally in July and August. The males of both species have a scent line in black

across each forewing. Both lay eggs on grasses that include Yorkshire fog, Timothy, cocksfoot or creeping soft grass. Consequently they are usually seen in the longer grasses at the northern end though both species will nectar on many plants, including bramble.

Essex skipper (top) showing the black underside of the antennae, compared to the small skipper below.

All species photos, unless otherwise indicated, are by Liz Cutting.

Large Skipper
This is indeed bigger than those previously mentioned, with an average wingspan of 33 to 35 mm. It is also on the wing before the others, my earliest record being 31st May, though not in the park. Similarities to the first two include a comparable range of grasses for egg-laying, having a single brood and, like the small skipper, overwintering as a caterpillar. It is not so moth-like in shape and the normal flight has less of the 'skipping' pattern. Size and earlier appearance help with identification, as does the fact that this is the only Suffolk skipper with orange and brown patterned uppersides to the wings. A freshly emerged adult is a beautiful sight. Its peak flight time is around the end of June and my first of 2016 was nectaring on bramble, close to the tennis courts, on 2nd. July.

In good breeding years skippers can become so abundant that it is not possible to accurately identify each one. This is why, on fixed route butterfly transect walks, where every sighting within a defined distance is recorded, both small and Essex skippers are listed on recording sheets as Sm/Essex. There is a fourth skipper in

Large skipper

Suffolk, the unfortunately named dingy skipper, but this is only found in West Suffolk, within small areas in and around the King's Forest.

Brimstone

This is one of the harbingers of spring, a large butterfly with the male a buttery yellow, having a spot, resembling mould, mid-wing. The female is a pale green that can at a distance create confusion with a large white. The average wingspan is 60 mm. Both the male and female settle with closed wings and with experience the wing shape distinguishes it from a large white. This is a species that hibernates in adult form, its closed wing shape easily mistaken for a dying leaf in brambles and ivy, the usual hibernating sites. The wings also have pointed tips, enabling water to drain off in hibernation, rather than freezing on the actual wings, which would be detrimental to survival. The male's colour is thought to give the 'butter coloured fly' description that may have been shortened to the word 'butterfly'. A new brood usually emerges in late July or August and these will eventually go into hibernation. This species needs buckthorn for egg-laying and, to hopefully improve numbers, a campaign was launched in 1998. This was masterminded by

Brimstone: note the pointed edge on the wing.

Julian Dowding from the Suffolk Branch of Butterfly Conservation. This county-wide campaign also had the support of Ipswich Organic Gardeners' group, Ipswich Borough Council and the Ipswich Wildlife group. Small 'whips' of buckthorn were used, with East Suffolk as the priority area. In two years about 2200 bushes were dug in at 200 sites, with gardens, allotments, golf courses and wildlife reserves among the areas concerned.

There is at present little buckthorn in the park, which probably is the explanation for only seven sightings between 2007 and 2015. However, there have been two sightings in 2016 and on 5th May my wife and I saw a brimstone just off the central path through the woodland reserve. It appeared to be egg-laying and by careful watching we managed to find one very small egg. As far as I am aware, this is the first record of brimstone egg-laying within the park and this will hopefully continue and increase once more buckthorn is introduced. This is discussed in more detail in the 'Future' section of this book.

Large and Small White

These can be easily confused, especially at a distance. Usually the large white is much bigger, from 63 to 70 mm, but some can be smaller, while small whites can often be quite large. The best differentiation is that large whites have a noticeable black border at

Large white

the top corner of the upperwing and this extends much further down the outer edge, compared to the small white. These are the two species that lay bottle-shaped eggs on cabbages and other types of crucifer though 'companion planting' of nasturtiums nearby may mitigate the effect. Nevertheless both species have great beauty

Green-veined white: text on following page

Small white

when seen flying through a blue sky. These are the two whites most likely to visit lavender beds but as they are mobile, with a powerful flight, they can be seen anywhere in the park and are one of the most common species.

Large numbers in some years may not be just the result of home breeding, with two to three broods each year, but also be caused by migration from the continent. In September 2000 I counted over 600 small whites flying in from the sea at Bawdsey beach. There is also an astonishing record, from a Norfolk botanical group, of an estimated six million large whites caught in insectivorous sundew at Sutton Broad. They overwinter as a pupa and can be seen in adult form from early spring to the beginning of November.

Green-veined White (see photo on p. 23)
This species is, in flight and by size, similar to the small white-both have a wingspan of about 48 mm. It is easily identified at rest by the pronounced 'veins' on the underwings, sometimes a green colour as the name suggests but usually a much darker shade. In some specimens this vein marking is just as pronounced on the upperwings. This is a more sedentary species compared to the previous two but, like them, is mobile within the park. It also overwinters as a pupa and though flying from early spring it is usually not seen so late as the other two. This is the 'good news' white butterfly since the eggs are not laid on cabbages or other crucifers. Instead it selects wild watercress, jack-by-the hedge, also known as garlic mustard, hedge mustard which grows at the top left corner of the park, or lady's smock. This last mentioned plant grows abundantly in the wet meadow and, as has already been mentioned, is slowly moving towards the nearby butterfly garden.

Orange Tip
In my early days as the Suffolk Butterfly Recorder I received a telephone call from someone in Trimley, asking me to identify a garden butterfly. The description given was 'white with orange tips' which was one of my easiest identifications. However, the female lacks the orange tips and can easily be mistaken for a similarly sized small or green-veined white. At rest the other difference is that both male and female underwings have a beautiful and intricate lichen-like patterning, in my opinion one of the most attractive of any British species. They have a less powerful flight than the other

Orange tip showing intricate patterning on the underwing.
See cover for upper wing

whites and the female usually 'flutters' along at a low height, looking for suitable plants on which to lay eggs. Those plants selected are as mentioned for the green-veined white. The eggs are usually laid singly on each plant, as the emerging caterpillars are cannibalistic. On 29th May 2016 and again on 4th June a female orange tip was observed egg-laying on hedge mustard, close to the copse at the park's northern end. The orange tip overwinters as a pupa and to me a first sighting is one of the great natural history moments that tell us spring has come, together with the song of chiffchaffs and returning swallows. The orange tip is usually flying between April and June. Being single brooded its breeding success can be affected by bad weather at this time of year.

Purple Hairstreak

Some years ago my wife and I were returning, in the evening, from Westerfield railway station. We quickly became aware that almost all of the large oaks bordering the road back to Ipswich were active with purple hairstreaks. This continued up to and including the large oak on the corner of Borrowdale Avenue, close to our home. Then there is a gap in oaks until the park is reached but my theory is that they were possibly deliberately planted and once stretched continuously from Westerfield village to the park, where oaks are abundant.

This small butterfly, with a wingspan of about 37 mm, is usually flying in July and is one of the most under-recorded and frustrating of our British species, since it is largely arboreal. It lays eggs in oak trees and it overwinters in this stage. The adults feed mainly on honeydew, a sticky liquid secreted by aphids and a substance known to most gardeners if they grow plants under or close to oak, ash or sycamore. In years when honeydew dries up, such as very hot summers, the butterfly will nectar at low levels, selecting such sources as bramble flowers, ragwort and buddleia.

It is a very attractive butterfly, with inky black upperwings which in sunlight are a deep purple, hence its name. Purple hairstreaks are most active in still and sunny conditions, with little movement

Purple Hairstreak (Peter Maddison)

when it is windy. Early morning and late afternoon into evening are the best times for observations and in late evening they are often active around the last sunlit leaves. Adults tend to patrol relatively small territories and if another butterfly passes there can be an aerial ballet which then involves others nearby. This sunlit activity has been compared to someone tossing up a handful of silver coins.

I recorded such activity on the evening of 24th July 2013, with sightings in six park oaks, my best so far. Watching purple hairstreaks for any length of time involves a degree of neck strain and my evening activities, complete with binoculars, were sometimes hindered by courting couples under some oaks. Binoculars are, however, essential, not just for distant sightings but because sometimes one active purple hairstreak can be traced back to a single leaf, where at rest its beauty can be observed in detail. There may even be a colony in the famous "Tawny Owl Oak" or nearby. On 8th August 2016 my wife found a weak and tattered specimen on the nearby path. Another on 23rd August was in a similar position near an oak close to the play area.

The white-letter-hairstreak can be seen in Suffolk but has not been recorded in the park.

Small Copper
This is one of my favourite butterflies, a small jewel with rich copper contrasting with black markings and borders. The undersides of its wings are very different, a brownish grey giving it very good camouflage when at rest on bare earth. It has an average wingspan of about 33 mm and is most frequently seen among the longer grasses towards the northern end, especially on yarrow. My best total in a small area has been six close to the frog seat. However, it has also been recorded within or close to the butterfly garden. It overwinters as a caterpillar and is normally first seen flying in April. In a good breeding year it can have three, even four, broods and my latest record, though not in the park, was on 16th November. Eggs are usually laid on sorrels and occasionally docks. For its relatively small size it is territorially aggressive and will chase off larger insects. Sometimes numbers recorded on one day can be incredible - 369 on a Butterfly Conservation survey of Rushmere Common and a count of 433 by my wife and myself at Tangham, part of the much larger Rendlesham Forest.

Small Copper

Most small coppers, with their wings open, are easily identified at a distance. A closer study may reveal the rare aberration *caeruleopunctata*, i.e. sky-blue spot markings. These can be as many as four just behind the copper band on the hindwings. Two have been found in the park: one in the Upper Arboretum and the other in the longer grasses at the northern end.

Brown Argus
It has only been recorded twice in the park, by two different recorders, in 2013 and 2014. Hopefully these two recent dates will support its continued existence. It is probably rare though identification often proves difficult as it can easily be confused with a female common blue, which has the same chocolate brown markings with orange crescents on the wing edges. The brown argus can, with experience, be identified by its silvery sheen in flight and lack of any blue colouring. Also at close quarters the spot marking on the forewing stops about halfway down to the body. In the common blue it continues right down. Incidentally close focus binoculars are very helpful in such situations.

Brown Argus

This butterfly colonised many new Suffolk areas during the 1995-1999 Millennium Survey, mainly due to a change in its egg-laying preferences - from rockrose, a rare plant in Suffolk, to varieties of storksbill and geranium, more widespread plants. This butterfly has two broods, from May to June and then August to September, occasionally into October. It overwinters as a caterpillar and, with an average wingspan of 29 mm, it is not easy to locate. Again, it is most likely to frequent the longer grass areas. The brown argus tends to have a jerky and erratic flight. Watch for where it lands, since it often returns to the same spot.

Common Blue

With a wingspan of about 35 mm, this is bigger than the brown argus and also overwinters as a caterpillar. As indicated the female is similar to a brown argus but males are a beautiful sky blue, both males and females having intricate underwing spot markings. This spot patterning in different colours is the best way of distinguishing this species from the holly blue. Eggs are usually laid on bird's-foot trefoil, black medick or restharrow and this is a species often best located late in the evening. They rest communally, heads downward, towards the top of grass clumps. This communal roosting produced a count of 208 on 8th August 2016, along a stretch of beach path between Thorpeness and Aldeburgh.

Common blues are not abundant in the park, with only eight records between 2007 and 2015, my best total being six on one day in August 2014. They are best seen from the highest path close to the tennis courts, looking down to the longer grasses. They usually have two broods, occasionally three, flying from late May into early October, normally with a July break between broods.

Mating Common Blues

Holly Blue

This usually has a more silvery blue colour and in my experience often flies at a higher level than common blues, though the species does use many lower nectar sources. Compared to the rich underwing spot markings on the common blue, the holly blue has a much simpler pattern, with small black dots and no orange marks. The sexes are similar and it has the same average wingspan as the common blue. However, it differs by overwintering as a pupa and consequently this is almost always the early blue butterfly seen in many gardens and in the park. A friend of mine has identified at least 28 different plants on which eggs have been laid, but the holly blue usually uses the fruit of holly for its first brood then ivy later in the year. Of the other plants used for egg-laying, a number can be found in the park, including buckthorn, buddleia, bramble, hawthorn, gorse and broom.

Holly Blue showing the difference in spot-marking compared to the more intricate design of the common blues opposite. The photo on the right shows 'mud puddling' which is more common in the 'blue' butterflies (see p. 32). Some species are also attracted to dead animals and faeces for reasons similar to those for 'mud puddling'.

It flies almost continually from April to September or early October and the holly blue is one of the species most likely to be seen 'mud puddling', that is using its uncoiled proboscis to suck up nutrients from wet earth, probably used by males to assist their mating activities. The holly blue also has a long recorded relationship with a parasitic ichneumon wasp, *Listrodomus nychemerus*, which injects its eggs into the holly blue caterpillar, with results that do not need to be described. The ebbing and flowing of this relationship obviously depends on the numbers of the host species and it is now recognised that there is usually a five year cycle. At its nadir holly blues are often just recorded at base colonies, at its zenith they are abundant. Many of the mature gardens close to or adjoining the park contain some of the plants used for egg-laying and it was abundant in 2016. In this year I recorded it from the garden hedge ivy next to the Soane Street entrance right up to the copse at the northern end of the park.

Red Admiral
I find this to be the most interesting of the park's butterflies. A few decades ago the sight of a red admiral in winter would warrant special mention in the annual report. Since then it has increasingly faced and survived our winters, to such a degree that in a recent 'Springwatch' programme it was described as a resident, rather than migrant, species. The best park example comes from 2014 when my wife spotted a red admiral on 23rd December. This was sunning itself on the wall of the visitor centre. On both 24th and 25th December what was almost certainly the same one had moved to the sunlit bricks on the front of Christchurch Mansion. Several photos were taken.

This is a big butterfly with an average wingspan of between 67 and 72 mm. Its vivid markings make it easy to identify with open wings and I consider it to be one of our most beautiful species. Even the more sober underwing patterning has a subtle close-up beauty and provides good camouflage. This is one of several park species laying eggs on nettles and one of the most likely to feed on fermenting juices of blackberries or orchard fruits. It readily nectars on buddleia and a good autumn spot for observation is at the end of the Park Road entrance. Here there are well established ivy clumps which offer a rich source of late nectar when their umbels open.

On 19th October 2014 I counted eight feeding here, alongside hornets and other insects.

Red Admiral feeding on a sap run

Painted Lady
It is amazing that such a seemingly fragile insect can migrate to our country from as far away as North Africa. However the birds, butterflies and other insects on migration are aware of propitious wind directions and thermals of air that will reduce the amount of energy they need to use. Migration numbers vary each year but 2009 was a memorable one. It started early in May and there was a large influx on the 24th. These butterflies were all looking worn and not stopping to feed, travelling north-west at about ten feet above the ground. Later they stopped to feed, laid eggs mainly on thistles and produced a subsequent summer brood. There were reports of mass movements all over the country, even from remote islands, and numbers were estimated in many millions.

This has a relevance to Christchurch Park as on the morning of the 24th my wife and I, entering the park from the Park Road gate, were quickly aware of these butterflies, moving as described above. They covered virtually all of the park from the Westerfield Road side to the Upper Arboretum but, not aware of their full significance, I didn't stop long enough to make a timed count - a pity, as this was probably the greatest influx of one species into the park in modern times.

Painted Lady

This species has a rich and colourful patterning on both the upper and underwings and is relatively large by British standards with an average wingspan of between 64 and 70 mm. Most eggs are laid on thistles but it sometimes uses nettles. It will nectar on buddleia and many other plants, having a short life cycle, sometimes little over a month from egg to adult. It is the only species endorsed by Butterfly Conservation as a kit to promote understanding of a butterfly's life cycle, especially for schools. Only recently has it been discovered that painted lady butterflies, which seldom survive our winters, do in fact attempt a reverse migration, back to the continent, in far greater numbers than previously thought. Observers have seen a few attempting this at relatively low heights but recent research has revealed that many travel back at much higher altitudes, presumably taking advantage of favourable thermals.

Small Tortoiseshell
In some years this species can be common, reinforced by others arriving from the continent. However, in recent years its numbers have declined, as a consequence of the actions of a parasitic fly, *Sturmia bella*, which injects an egg into the caterpillar and on emergence it proceeds to eat its host. The effect on population numbers was considerable for several years, both nationally and in the park. The small tortoiseshell is an early butterfly, coming out of hibernation and in 2009 seen by me on 15th March. Then the problems began and there were no park records for the next three years. Recovery began in 2013 but the first sighting wasn't until 30th June. Its recovery since then is shown by the last three years having first sightings as follows: 2014-24th February; 2015-8th March; 2016-10th March.

With an average wingspan of 50-56 mm, this species is appreciably smaller than the red admiral and it also lays eggs on nettles. These hibernating butterflies can with good fortune have a lifespan approaching a whole year and those emerging from long months of overwintering often appear in surprisingly pristine condition. This species is the most likely to hibernate inside a building, be it house, church, shed or other outbuilding. Despite its ubiquity in good breeding years, the small tortoiseshell is definitely worth a close look - the blue crescents along the outer edges of the upperwings

Small tortoiseshell

are particularly beautiful. As with many colourful species the underwing patterning is intended to offer good camouflage especially on bark or bare earth. During the 1995-1999 Millennium Survey this was the second highest species across Suffolk, in terms of numbers of recording tetrads in which it was present.

Numbers are considerably increased in some years by large movements from the continent. Hungry for nectar, these will on reaching land congregate en masse on any available nectar source, such as the clumps of valerian growing in the shingle on Aldeburgh beach. Once fed they tend to disperse as they move inland but I did count 122 on 20th June 2014, exactly half being in one field. This was in the Fynn valley, only about four miles from the park. Regrettably my best count within the park has been just six.

Peacock
In February 2014 Reg Snook invited me to join him in an exploration of two World War Two 'pillboxes' not far from his studio at Grundisburgh. Sixty hibernating butterflies were found and the 'pillboxes' seemed ideal for hibernating butterflies, with an equable temperature, very dark corners, minimal light coming through the narrow windows and virtually no disturbance. Here two species were hibernating, small tortoiseshells and the majority being peacocks. Even in the limited light afforded by a torch the

Peacock (see frontispiece for upper wings)

large wing rings of the peacocks were visible, although their wings were closed. These beautiful and large rings explain the name and they are also used to good effect if a predator arrives during hibernation. By communal hibernation they can open their wings in a synchronised manner and this multitude of large flashing eyes can deter many predators. The other defence mechanism is an impressive noise made by their rustling wings, giving an audible hissing which with the eye flashing and relatively large wingspan of about 66 mm obviously helps to survive the long months of hibernation. So far, I have failed to find any communal hibernation within the park, though the ice house could be suitable.

This is another hibernating species that lays its eggs on nettles. The summer brood is usually flying from late July. A newly emerged summer peacock has a superb 'velvet' sheen to its colourful wings. The underwings are very dark to give it good camouflage and a peacock in flight is the only British butterfly to have a coal black appearance. For some still unknown reason many peacocks go early into hibernation - there is only one park record after August.

The peacock has the most spectacular eyespots of any British butterfly. Many others also have eyespots, invariably at the wing edges. These can distract predators from attacking the butterfly's more central vital organs, whereas a butterfly with damaged wingtips can normally survive and carry on flying.

Comma | Underwing showing the distinctive white mark that gives the species its name.

Comma
This is the last of the five butterflies hibernating as adults. Its recovery is remarkable as at the start of the twentieth century it was rare and absent from most areas. A change of plant on which it lays eggs, from hop to nettle, led to a recovery with the comma still slowly spreading northwards.

This species has an average wingspan of 55-60 mm and is the one of the five most likely to hibernate in a more exposed outdoor location, usually the sheltered side of a tree branch or trunk. The notched and indented wings make it unique among British butterflies and at rest the bark-like underwings, with the white comma mark, make it look remarkably like a dead leaf. Once, I located a comma with closed wings in a relatively large tree, only noted because I saw where it landed. Then I deliberately took a distant photo to show just how well it was camouflaged. With wings open it is a tawny orange with black markings and the shades of colour can vary from light to dark. In flight it has an easily identified orange colour and, in Suffolk, can only be confused with a flying silver-washed fritillary, a much larger species which has not been recorded in the park. Compared to most peacocks it is on the wing over more months of the year: park records are from 4th March, emerging from hibernation, to 29th October, with this next brood close to hibernation.

Speckled Wood

On 13th May 2001, the day after the first speckled wood was recorded in our Westerfield Road garden, we deliberately went to look for it in the woodland reserve and found two close to the sunlit central path. These were the park's first records. This butterfly has had a remarkable colonisation of Suffolk. When I first came to Ipswich in 1966 it could only be found in the Brecklands of West Suffolk. Now it has spread widely, only rare in a few areas such as the Waveney valley from Diss towards the sea. This is because there is a lack of habitats with the dappled shade it prefers. Edges of woodland are where it can often be found. One such park area is along the rough path from the orchard gate downwards, which follows the woodland edge. It can be seen elsewhere in the park, including the northern copse.

One possible reason for its success is its unique ability, compared to any other British species, to overwinter either as a caterpillar or pupa. If the latter occurs then the adult will emerge earlier in spring. This dual emergence gives it two different times for breeding success and therefore more opportunity to beat the vagaries of spring weather, compared to the single emergence of another spring butterfly, the orange tip. With an average wingspan of 47-50 mm this species can be flying from March through to October: park records are from 4th April to 23rd September. Eggs are usually laid on grass blades under shrubs and the underwing pattern of greys and browns gives it effective camouflage at rest. The upperwings are chocolate brown with creamy yellow patches, a beautiful mix which closely replicates dappled shade. If you see two butterflies flying together in the woodland reserve they will almost certainly be speckled woods. The males vigilantly guard their territory, especially favoured sunspots, and vigorously challenge any others that intrude. Detailed research in an Oxford wood has proved that in such territorial battles the resident butterfly almost always wins.

Dappled shade on the path by the side of the woodland reserve, which is the preferred habitat for speckled wood.
(Habitat photo: Richard Stewart)

Wall or Wall Brown

Unfortunately we have almost certainly lost this species from the park. The only record is mine from 11th October 2003. It has suffered a major decline nationally and despite much research the causes are still not clear. It is now largely restricted to Suffolk coastal sites and in 2015 was only recorded from 22 recording tetrads, compared to well over 300 during the 1995-1999 Millennium Survey. At this time it was still seen regularly inland. However it was last recorded from my Fynn valley butterfly walk transect back in 2004. As already indicated, the reasons for its steep decline are not clear, as it lays eggs on a variety of common grasses and can be seen from late April right through to October, with the possibility of a third brood in exceptional years. It overwinters as a caterpillar and is slightly smaller than the speckled wood. It is well camouflaged on its underwings but the upperwings have a bright orange base, with darker borders and a range of different sized eyespots on the wingtips and around the lower edge of the hindwings. The wall also has the habit of flying just a few yards in front of a walker, then repeating the process once you are too close. It can often be found basking in sunlit areas of bare earth.

Wall or wall brown.

Grayling

Again there is just one park record. On 14th August 2009 my wife Anne-Marie spotted one perched on a seat close to the Soane Street entrance. This is not a butterfly associated with any habitats in the park, since it is usually found in the Sandlings heathland, with large colonies within some coastal sites. However we have recorded it on our garden buddleia and this is a deceptively large butterfly with a wingspan of between 55 and 60 mm. This is effectively reduced to 32mm, since it always has its wings closed after landing. The one exception is during courtship, one of the most elaborate of any British butterfly. My wife has studied this, especially at the large colony in and around the dunes at Winterton on the East Norfolk coast. Her published article was titled 'Bow To Your Lady'. This elaborate courtship is unlikely to be seen in the park as the grayling's required grasses on which to lay eggs are not present.

At rest its hindwings are a marbled pattern of browns and greys and it is a master of camouflage on bare soil. Not only does it tuck in the conspicuous eyespot but it can also tilt its body to reduce the amount of cast shadow. Like the wall butterfly it often flies just a few yards ahead, repeating this when someone gets too close. Having said that, it also has a behaviour not usually included in butterfly books, which is landing on the flesh or clothes of humans, presumably to suck up salts and other nutrients, particularly on hot days of perspiration. Overwintering as a caterpillar, this is a butterfly of high summer, though its flight period can extend further into the year - my latest record is for 28th September. There is no heathland in the park but this habitat still survives in relic areas on the edge of Ipswich, from which the Sandlings once spread through to Southwold. Grayling colonies do still exist in these smaller heathlands and at Purdis Heath good management work has increased their numbers. Given the right wind direction the grayling could be seen again within the park, as it will readily feed on nectar sources such as bramble and buddleia.

Gatekeeper or Hedge brown

The two names probably originate from this species frequenting hedges and flying up when disturbed from gates at woodland edges. It usually emerges a few weeks later than the meadow brown or ringlet, which will be described later. This is another butterfly of

Grayling

Bottom photo shows the grayling's predilection for landing, often repeatedly, on human clothes or skin. It tends to prefer brighter colours, especially floral patterns. (Bottom photo: Richard Stewart)

high summer, usually flying from the second half of July to the end of August and peaking in abundance during the first half of August. It is smaller than the other two species mentioned, with a wingspan averaging from 40 to 47 mm. Females are usually larger and males have a conspicuous dark scent gland on the forewing. The other difference is that this species is a more golden colour with large orange areas enclosed by grey and brown borders. In a good breeding year gatekeepers can be abundant and it is also one of the few summer butterflies to rest with wings open in cloudy conditions. It overwinters as a caterpillar and lays its eggs on a variety of grasses.

Gatekeeper

Gatekeeper underwing

Meadow Brown
This is suitably named, being a species abundant in summer meadows and largely brown in colour, though with eyespots and areas of orange. In the park it is usually easy to see in all the areas of longer grasses from the end of the wet meadow up to the northern edges. However, this butterfly may be seen all over the park, feeding on a wide variety of nectar sources, especially buddleia and hawkweed. It overwinters as a caterpillar, has a wingspan of 53 mm and the eggs are sometimes simply dropped on a variety of wild grasses. It was recorded from more tetrads than any other Suffolk species during the 1995-1999 Millennium Survey and a detailed count in the park would almost certainly prove it to be the most abundant summer butterfly.

Meadow brown and a mating pair (bottom).

Ringlet

This is usually slightly smaller than the meadow brown but it also overwinters as a caterpillar and lays eggs on a range of wild grasses. It tends to be on the wing just after the meadow brown and before the gatekeeper. The name derives from the delicate ring markings on both the upper and underwings, though those on its open wings are sometimes difficult to find. It tends to have colonies in slightly damper habitats though the best site in the park is along the steep bank behind the former basketball court, number 7 on the habitat list. It usually has a short flight period from July into the first half of August. In flight it is much darker than the meadow brown, lacking the lighter shades. At close quarters there is a silvery white fringe around the outer wing edges. Compared to both gatekeeper and meadow brown this is a much rarer park species.

Ringlet nectaring on bramble, which is abundant in the park and one of the most important nectar sources for summer insects.

THE FUTURE

This park has an abundance of the two main summer nectar sources. There are extensive beds of bramble and the buddleias include large flowering groups on the islands in the Wilderness Pond. Buckthorn, the only plant on which the brimstone lays its eggs, needs to be increased and this is in hand with the purchase of several hundred which will hopefully be planted in and around hedges of native species. In addition there will be more areas of gorse and broom, plus honeysuckle in the woodland reserve. I have also suggested the planting of two herbs, marjoram and thyme, both native varieties that are particularly good for bees and many other insects, including a wide range of butterflies. Aubretia, alyssum and arabis will be added as early nectar sources in the butterfly garden. This additional planting may help to attract some of the species listed later, but not presently recorded in the park.

Marjoram on the right of the picture with thyme to the left: both will feature more prominently in future park planting. They attract a wide variety of bees and butterflies. (Photo: Richard Stewart)

Swallowtail

There are three tantalising records from close to the park but none actually inside. I recorded one passing through our garden on 28th September 1998 and in the same month I received a report of a caterpillar in the garden of Mrs. Spooner, literally about fifty yards from the park. This was feeding on garden rue but did not survive. Since the British swallowtail, subspecies *britannicus*, feeds as a caterpillar almost exclusively on milk parsley, it was almost certainly the continental subspecies *gorganus*. This has a much wider range of plants on which it lays eggs, including rue. Both subspecies are virtually identical and later investigations revealed an almost certain covert breeding and releasing from someone in Ipswich - one actually flew past the window of what was then the Borough Council offices near the Wolsey theatre. Since the swallowtail is one of a few British butterflies to be legally protected in all stages of its life, this action was a criminal offence. The third sighting was by Pat Gondris on 4th August 2004, at Saint Edmund's Road, again close to the park.

Swallowtail: showing the 'tails' which help to explain its name.

The swallowtail, our largest and most spectacular British butterfly, has a wingspan of 80-90 mm and can be seen in the Norfolk Broads from late April through to the second brood on the wing until mid-September. Occasional Suffolk sightings from the Waveney valley could be wandering *britannicus* from the Broads. They overwinter as a pupa and any park sightings would be 'passing through', though possibly lingering to feed on nectar sources such as bramble and buddleia. The continental subspecies does migrate so any local sightings could be of genuine wild butterflies. Often such migrants, moving inland from the coast, follow natural features such as rivers. This large butterfly, with a powerful and gliding flight, could easily cross from the Orwell to Christchurch Park.

Swallowtail, showing the underwing patterning. On a sunny day the effect in such a position is reminiscent of a stained glass window.

Clouded Yellow

This migratory species is rarer than the painted lady and less likely to be seen inland. At a distance it could be mistaken for the larger male brimstone but is usually a deeper yellow, though lighter specimens do occur. There is a black spot halfway across the forewing and a pair of silver spots, surrounded by reddish brown, on the hindwing. Any arriving on our shores will produce a second and possibly a third brood, usually laying eggs on the uppersides of leguminous plants such as clover and lucerne. It has occasionally overwintered successfully in very sheltered southern sites but never in Suffolk. Almost all of my records come from coastal sites, especially a 'hotspot' such as the coastal walk from East Lane, Bawdsey towards Shingle Street. However, as with the painted lady, numbers fluctuate each year.

Clouded yellow

Green Hairstreak

I am surprised that this has not been recorded in the park until May 2016 since it has been seen in many parts of Ipswich. There are early nectar sources in the park it could use, such as flowers of hawthorn and rowan. Its spread has probably been along the 'green corridor' of the Ipswich to Felixstowe line, where there is plenty of gorse and broom, the two plants on which it lays eggs in Suffolk.

The lack of records could be explained by one or more of these reasons - first it is a small species, with a wingspan averaging about 33 mm and effectively reduced to 15 as it always rests with wings closed. Overwintering as a pupa it emerges early, usually from April through to June, at a time when not so many butterfly watchers are around. Third, although it is our only British green butterfly, it is well camouflaged against vegetation and looks brown in flight. The final reason is that there is only a limited amount of gorse and broom in the park, mainly close to the tennis courts. That is why both feature in the additional planting.

Green hairstreak: many butterflies have beautiful antennae.

Postscript: A record of this species was given to me by Adrian Richards, seen on 14th May 2016 below the tennis courts, where there is some of its food plant. I was informed of this record too late for it to be included in the main species text and charts, which had already been finalised.

White Admiral

This is a beautiful butterfly, with a dark brown upperwing and conspicuous white bands. To me the underwing patterning of whites, bronzes and blacks makes it one of our most beautiful species, its grace and beauty further enhanced by a superb gliding flight. It has a wingspan of 60-64 mm and overwinters as a caterpillar. It is essentially a high summer species, usually on the wing from late June until the end of July. It has slowly colonised local areas, being recorded from Wherstead Wood and, in 2015, Jan Cawston took a photo of one at the Dales, little more than a mile from the park. In 2016, two were recorded at Nacton. It lays eggs on honeysuckle but needs the plant to be in at least partial shade, not luxuriant sunlit clumps. There is one area of honeysuckle at the north end of the woodland reserve and more will be planted in suitable areas.

White Admiral

White Admiral: the subtle underwing pattern.

Silver-Washed Fritillary

I am in ten different branches of Butterfly Conservation and have, by reading their different newsletters, been able to monitor the recent spread of this species. Our largest fritillary, with a wingspan of 72-76 mm, is spreading through Suffolk. On 15th July 2015 one was in our garden and obligingly made a second circuit so my wife could also see it. It fed briefly on buddleia and will readily nectar on a variety of plants. It initially lays eggs on trees and the emerging caterpillars crawl to violet plants for feeding.

It overwinters as a caterpillar but I am not aware of many violets in the park. There was a good clump a few years ago, close to the town end entrance to the woodland reserve, but they were subsequently covered by brambles. This is a powerful flier and could easily reach the park from its recently colonised areas.

Silver-washed fritillary: the name is explained by the underwing patterning resembling waves breaking at an angle on a beach.

PREDATORS IN THE PARK

As Reg Snook has often mentioned in his park wildlife notes, there is continual predation. The most obvious is the catching and killing of young waterbirds by large gulls. Also conspicuous is the plucking and devouring of prey by sparrowhawks, which nest in the park, at the northern edge of the woodland reserve. The predation of butterflies in their different stages is less obvious, mainly because they are much smaller. In the text for holly blue and small tortoiseshell there is mention of species specific parasites laying eggs which hatch inside the caterpillar and then devour it. This type of predation can occur in other butterfly species as well. Fungal infections may afflict any of the developmental stages and eggs can be eaten before they hatch. Caterpillars are taken in their thousands by birds, especially when young nestlings need to be fed. This explains why butterflies need to lay so many eggs, in order that enough survive and the caterpillars are in sufficient numbers to reach their next stage of development. Adult butterflies are also often caught in flight by many species of birds.

The hornets, which are particularly noticeable around the open nectar-bearing umbels of ivy in autumn, are also after prey. If they catch a butterfly the wings will be snipped off before it is eaten. Many years ago, before redevelopment at the RSPB's Minsmere reserve on the Suffolk coast, there was a sunlit buddleia close to the toilet block. Here hornets regularly hunted and caught butterflies, even the larger species. Their success could be measured by the many discarded butterfly wings beneath the bush.

Butterflies on or close to the ground can be caught by predators, including the small mammals living in the park. Adult butterflies are particularly vulnerable during early mornings of heavy dew, making flight difficult. Also a freshly emerged adult has to survive several hours during which the wings dry out and are sufficiently strengthened for flight. Other predators include debilitating mites which attach themselves to the adult's body. Spiders also make their webs across much visited nectar sources such as bramble and buddleia. Given its forward momentum towards a nectar source, and the speed at which a spider tackles its caught prey, even a butterfly as large as a red admiral has little chance of escape.

Crab spiders (see photo) can camouflage themselves superbly on plants, waiting until the feeding or egg-laying butterfly ventures too close. With names such as 'darter', 'hawker' and 'chaser' dragonflies have an amazing acrobatic ability to catch prey, further aided by their large compound eyes (see photos overleaf). As with hornets, they also will remove the wings before the prey is eaten. I have witnessed this in our garden at Westerfield Road and in the park the numbers of both dragonflies and damselflies have increased recently. They are obviously attracted to the water but more recently Canadian pondweed has been another factor. This now covers much of the Round pond and is used by them for egg-laying.

A crab spider has caught this small white.

Adult butterflies in hibernation are also vulnerable and this explains why they often do this communally and have patterns and colours on their closed wings that improve their camouflage. Other hibernation 'defences' are included in the peacock text.

Perhaps the greatest indirect, but sometimes devastating, predation is from human interference. In 2016 I could find only one good butterfly habitat that suffered from premature cutting, number 5 on the map, and it is good to report that all the habitats listed are still intact. Joe Underwood, the Education and Wildlife Ranger, has kept a close eye on these areas. This is not as straightforward as it may sound, since there are many other interest bodies connected to the park, as well as the wildlife one.

Large skipper: one defence against predation is that almost all species have a more subdued underwing pattern and colour.

Two of the dragonflies that predate on butterflies: female broad-bodied chaser (top, Richard Stewart) and female black-tailed skimmer (bottom, Anne-Marie Stewart)

Summary of Park Butterfly Species - in alphabetical order
There are 59 species either resident in Britain or regularly migrating here. Of these 33 can be seen in Suffolk in a naturally occurring condition. The following Suffolk species are unlikely to ever visit the park because of their sedentary nature or specialist habitat requirements:

Dingy skipper   Silver-studded blue
Small heath   White-letter hairstreak

The following, as indicated in the text, could be seen at a future date:
Clouded yellow   Silver-washed fritillary
Swallowtail   White Admiral

These three species have only been seen once in the park since 2003: Grayling, Green Hairstreak, Wall or Wall brown

These are regularly seen:
Brimstone   Brown argus
Comma   Common blue
Essex skipper   Gatekeeper or hedge brown
Green-veined white   Holly blue
Large skipper   Large white
Meadow brown   Orange tip
Painted lady   Peacock
Purple hairstreak   Red admiral
Ringlet   Small copper
Small skipper   Small tortoiseshell
Small white   Speckled wood

A final caution: unusual, exotic butterfly species sometimes escape from the butterfly house at Jimmy's Farm in Wherstead and they could reach the park. Species not naturally occurring in Suffolk could also arrive as a result of clandestine 'breed and release' activities. These are at present largely confined to the Piper's Vale and Landseer Park areas. Butterfly Conservation strongly disapproves of such activities and was one of the pioneers in formulating a detailed code for reintroducing butterfly species. This code has been adopted and largely implemented by all the major national wildlife and conservation groups.

The brimstone, being a butterfly that hibernates in adult form, is one of the longest living species. It has been recorded nationally in every month of the year.

Summary of the Flight Period of Butterflies Recorded in the Park

This covers the 24 species listed in this book, in alphabetical order. The normal flight period in Suffolk is shown with the first and last recorded dates in the park, between 2003 and August 2016. Where there is a significant difference, for example with Essex skipper, grayling and wall brown, this is the result of only a few park records being available.

Park earliest and latest dates.

Brimstone: 8th March; 4th August.
Brown argus: 25th May; 8th August.
Comma: 4th March; 29th October.
Common blue: 25th May; 20th August.
Essex skipper: 13th July; 6th August.
Gatekeeper: 12th July; 16th August.
Grayling: 14th August- just one record.
Green-veined white: 30th March; 22nd September.
Holly blue: 9th April;4th October.
Large skipper: 17th June; 29th July.
Large white: 9th April; 9th September.
Meadow Brown: 6th June; 12th August.
Orange tip: 16th April; 3rd June.
Painted lady: 24th May; 5th September.
Peacock: 8th March; 5th October.
Purple hairstreak: 5th July; 23rd August.
Red admiral: 8th March; 25th December.
Ringlet: 7th July; 2nd August.
Small copper: 6th May; 11th October.
Small skipper: 30th June; 2nd August.
Small tortoiseshell: 24th February; 20th August.
Small white:2nd April; 7th September.
Speckled wood: 4th April; 23rd September.
Wall brown: 11th October- just one record.

Species	JAN	FEB	MAR	APR	MAY	JUN	JUL	AUG	SEP	OCT	NOV	DEC
Brimstone	7		8					4			27	
Brown argus		15		24	25			8		27		
Comma			4							29	6	12
Common blue				7								
Essex skipper					25		13	20				
Gatekeeper					29		12	16	12	6		
Grayling						10				1		
Green-veined white			11 30			21		14	22		5	
Holly Blue			18	9						4	13	
Large skipper					21	17	29		18			
Large white			18	9								
Meadow brown					5	6		12	9	6		25
Orange tip			23	16		3	20					
Painted lady		20			24				5	5	20	
Peacock	1		8			13	5	23	8			31
Purple hairstreak							7	27				
Red admiral	1		8			3	2					25 31
Ringlet				6		30			25	11	16	
Small copper			20		17			2				
Small skipper								20				
Small tortoiseshell	1	24		2					7		19	28
Small white		21		4					23		16	
Speckled wood			27	1						11 30		
Wall												

Legend (Suffolk)

5 Earliest   Occasional   MAIN FLIGHT   Occasional   Latest 5

Earliest and latest dates for Christchurch Park are shown in Blue and in black for Suffolk, up to the end of 2012.

CREATING A BUTTERFLY GARDEN

As Reg Snook has often remarked in his park wildlife notes, the birdlife is enriched by the fact that many large and mature gardens are close by, where birds are not only fed regularly but also recorded. This can also apply to butterflies. Anyone with a relatively sunny garden, irrespective of size, can attract butterflies by planting good nectar sources. The provision of plants suitable for egg-laying can mean that the whole life cycle of a butterfly could take place in one garden. At this point can I add a word of caution about nettles? They are needed by at least four species for egg-laying but seldom succeed in gardens. Butterflies just seem to prefer them elsewhere.

Try to select native plant species and avoid double-flowered varieties, which seldom attract many butterflies. Buddleia is of course an exception, not a native plant but now also known as the 'butterfly bush'. A member of Butterfly Conservation, admittedly with a very large garden, has over thirty species which between them flower for most of the year. Sunlit clumps are better than single specimens and it is also helpful to look for the 'butterfly friendly' symbol which should be in good nurseries. The ultimate aim is to provide plants to attract butterflies from hibernation emergence right through to autumn. In our garden we have recorded 28 species, using over 50 nectaring plants. Obviously any use of insecticides will have a negative effect as will excessive removal of 'weeds'. The following list is not all embracing but gives a seasonal guide to proven nectar sources.

Spring - aubretia, alyssum and arabis, dandelion, hyacinth, flowers of laurel, cotoneaster and pyracantha, Mexican orange blossom, the early flowering *alternifolia* buddleia and some plants used for egg-laying such as honesty, garlic mustard, lady's smock and buckthorn.

Summer - various and mainly purple forms of *davidii* buddleia, pruned back early in spring and regularly deadheaded to prolong the flowering period. This will probably be the most successful garden nectar source, and may also attract the spectacular hummingbird hawk moth. Thyme and marjoram attract many bees and butterflies but select native varieties. *Verbena bonariensis* is tall but so shaped that it can be used at the front of a border and has a

View of the park's butterfly garden. Beyond the seat contiguous habitats link up the wet meadow and longer grasses towards the top of the park. (Photo: Richard Stewart)

long flowering period. The wallflower Bowles' mauve also has a long period of flowering as do erigeron and valerian. Other good summer nectar sources include golden rod, phlox, sweet William, candy tuft, lavender and dames' violet.

Autumn - fallen soft fruit, Michaelmas daisies, preferably purple varieties, *Sedum spectabile* rather than the autumn joy variety and late ivy umbels for feeding insects, hibernation sites and birds feeding on the winter berries. Try also to include the *weyeriana* variety of buddleia, late flowering which, with regular deadheading, can still be producing new flowers well into winter.

Your garden should also provide hibernating sites but be careful with species such as the small tortoiseshell, often hibernating in sheds, other outhouses or indoors. If they emerge very early, as the consequence of a sunny winter's day, resist the 'catch and release' temptation as they will probably die outside. It is best to gently catch the butterfly in a large container such as a box, and then return it to a cool, shaded place with a relatively equable temperature.

If you are reading this and becoming conscious of your own garden limitations, remember that many of the above plants, including some species of buddleia, can be grown in pots or other containers.

Nectar sources attract butterflies into gardens but an even greater satisfaction can be had from having the whole of their life cycle taking place in your garden. This can be achieved by providing the sources for egg-laying already indicated in the species section. I have already expressed reservations about garden nettles but holly and ivy will attract egg-laying holly blues, a species closely associated with confined habitats such as churchyards and gardens. They prefer native species of these plants, for holly *Ilex aquifolium* and for ivy *Hedera helix*. However, I have seen ornamental species also being used. Both green-veined whites and orange tips lay eggs on garlic mustard, which in a garden should be kept in one area as it can become invasive. Orange tips also use it for roosting.

Bird's-foot trefoil can attract egg-laying common blues and this plant is also suited to containers. Honeysuckle, grown partially in shade, may encourage the rarer white admirals. Buckthorn, the

Top: some of the buddleia collection in Trudie Willis' garden (see p. 73).
Bottom: valerian in the author's garden. (Richard Stewart)
Both buddleia and valerian also attract many moth species.

brimstone's only egg-laying source, produced an interesting series of events in our garden. Nine caterpillars were found, despite their excellent disguise, but these were eaten by the alien harlequin ladybird. However, one survived, as a freshly emerged female was seen nectaring on our garden buddleia.

Buckthorn can also be readily incorporated in another garden feature, a hedge of native species. This, if successfully planted, can attract a wide variety of wildlife including, once it is thick enough, nesting birds. Different stages of butterfly development, including hibernation, can occur here and in the park there are three examples: by the side of the Reg Driver visitor centre, inside the orchard and at the northern end of the park, towards the Westerfield Road fence. Each have gaps that will hopefully be filled in future with more buckthorn. Such hedges can include a rich mix of native species such as hawthorn, field maple, ivy, buckthorn, blackthorn, spindle, dogwood, guelder rose, hazel and holly. Several of these offer winter berries and colour, especially holly, spindle and dogwood.

Such a hedge can also be an effective way of dividing a wilder area from more formal garden planting and design. This introduces the concept of the wildflower meadow. The easiest way to do this is simply by letting part or all of a lawn revert to nature and remain uncut until late summer, just to see what comes up. There is a story, hopefully not apocryphal, of a man suddenly rushed to hospital. He was there for several weeks and on returning home he found his overgrown lawn now included bee orchids. Often nature is just waiting for a chance.

Such a 'meadow' can be relatively small. The close-up photo on page 71 is of an area no larger than ten metres by five metres, unusually in a front garden, on the corner of Moat Farm Close in Ipswich. These wildflower meadows will of course attract more wildlife besides butterflies and could also incorporate 'hedgehog homes', 'insect hotels' of bamboo canes and easily constructed log piles to help one of our most impressive insects, the stag beetle. There are several of these piles in the park. The area selected should be in a sunny spot and such a project, to be successful, needs soil that is not too high in fertility. You may need to strip off some topsoil. Also avoid the use of rye grass. This is ideal for tough-

The bottom photo shows a brimstone caterpillar well camouflaged along a spine of buckthorn leaf. The picture above shows a caterpillar having devoured most of the leaf (Anne-Marie Stewart).

wearing lawns but in a wildflower meadow it will encourage vigorous and dominating grasses at the expense of flowers. As well as the grasses already mentioned for egg-laying in the species texts, some of these should also be included: brown and fine bent, red and sheep's fescue, rough and smooth meadow grass, plus others with wonderful names such as meadow foxtail, sweet vernal-grass, wavy hair-grass and crested dog's-tail. Also add the beautiful quaking grass.

The easiest way to proceed is to purchase a commercial wildflower seed mix, which should always be of native, and preferably local, origin. Try to include these three flowers: yellow rattle - yes, the seeds actually do rattle - which helps to reduce the effects of more vigorous grasses by attaching itself to their roots, from which it extracts water and minerals. The other two, corn marigold and corn cockle, as part of their name suggests, were once familiar plants within or on the edges of cultivated fields.

A final important point to remember is not cutting prematurely: leave at least until the end of summer, to enable both plants and insects to complete their life cycles. Cuttings should then be removed to avoid enriching the soil.

However, narrow paths can be made and regularly mowed to allow access and greater enjoyment. The long grass areas to the north of the park have such paths, especially below the tennis courts and leading up to the frog seat. Some of these defined areas could, with the use of a suitable seed mix, be converted into such wildflower meadows.

The Suffolk Wildlife Trust has helpful leaflets - see the later information - and its reserves handbook has at least twelve places that are either simply wildflower meadows or include such a habitat alongside others. Three are particularly recommended for an inspirational visit as they cover the three seasons of growth. The first is Mickfield Meadow, with spring flowering snake's head fritillaries and goldilocks buttercups, then in early June at Nacton Meadows there is ragged robin and an abundance of southern marsh orchids. In early September Martin's Meadows at Monewden host the meadow saffron, growing in fields bordered by some of Suffolk's oldest, most impressive and species rich hedges.

(Richard Stewart)

Top picture: some of the orchard area could also be made into a wildflower meadow. Bottom picture: the front garden 'meadow' mentioned in the text. It includes a good range of nectar sources many of which have largely disappeared from cultivated fields.
(Photos: Richard Stewart)

Several local nature reserves, especially on the southern and western edges of Ipswich, also have grasslands with a good variety of wild flowers. There is a particularly good example of species-rich grassland at Landseer Park and The Dales, only about a mile from the park, has a large sunlit meadow that is increasing in wildflower species thanks to successful management and some additional sowing of seeds. At their very best such wildflower meadows can have as many as thirty different plant species in a single square metre. That rich concentration is highly unlikely in a garden version but any contribution, however limited in size, will help to correct the loss of 95% of this habitat that has occurred during the last forty years.

As a perhaps unexpected bonus, moths will also be attracted to the nectar sources. Many moths do in fact fly during the day and the list includes the spectacular hummingbird hawk moth, which can still be feeding almost at dusk . In our own garden it has been recorded on buddleia, *Verbena bonariensis*, summer jasmine, valerian and Bowles' mauve wallflower. This is a migrant, occasionally overwintering in a few sheltered southerly areas. During one feeding session on valerian I recorded it visiting fifty different florets in just thirty seconds. This incredible activity explains the need for constant feeding to replace the energy used. The silver-y is also a migrant, feeding by day and into the night on many different nectar sources, especially buddleia. In 2009, as the text for the painted lady indicates, millions of these butterflies came over but their numbers were probably exceeded by silver-y migrants. This species is named after the prominent wing marking.

The fact that moths are attracted to these garden nectar sources may help to dispel the still commonly held belief that all are small, drab and likely to ruin your clothes. In reality very few are harmful to clothes and many of the thousands of species have an amazing array of colours and patterns, especially the larger hawk moths.

As well as visiting the park's butterfly garden for planting ideas, two other gardens are well worth a visit, admittedly some distance from Ipswich. The first is the RSPB's wildlife garden at Flatford. This was started in 2011 and has the convenience of being close to the car park. It also benefits from being in a sheltered position and

has a wide variety of nectar sources plus an area of longer grasses for egg-laying. Here you can also watch the activities of leaf-cutter bees while its proximity to the River Stour attracts a wide variety of dragonflies and damselflies.
Details: rspb.org.uk/flatford or by phone at 01206-391153.

The second is close to the Suffolk coast at Prior's Oak, along the Aldringham Road at the edge of Aldeburgh. Here Butterfly Conservation member Trudie Willis has created a wildlife haven that extends to ten acres. Habitats range from more formal lawn and vegetable areas right up to the wilder area at the top of the garden. Being close to the RSPB's North Warren reserve it attracts some species not seen in Christchurch Park, such as white-letter hairstreak and small heath. It is also a good place to see two species relatively rare in the park, namely grayling and the migratory painted lady. Both will probably be feeding on one of large collection of buddleias in the garden, these having a wide range of colours and flowering periods. Dragonflies, damselflies and lizards by the pond add an extra element, there are nesting swallows and purple hairstreaks in the oaks where visitors park their cars. Muntjac deer, foxes and badgers have also been recorded.

Details: open for the benefit of a range of charities between June and September, including a Butterfly Conservation day that has guided walks. It is also open by private arrangement.

Butterfly Conservation has a leaflet about butterfly gardening, from www.butterfly-conservation.org (click on 'How You Can Help' and then 'Gardening') or telephone 01929-400209.

The Suffolk Wildlife Trust can post, or send to your e-mail, leaflets about grasslands and wildlife gardening. Tel: 01473 890089 or info@suffolkwildlifetrust.org.

Any butterfly records from the park or any other Suffolk sites can be sent, with name, address or e-mail plus the date and location, with preferably a grid reference or postcode, to: butterfly@sns.org.uk.

EPILOGUE
Haiku- Four Seasons Of Butterflies In Christchurch Park

Beneath a blanket
Of soft snow on deep ivy
Yellow brimstones rest.

Earth is still waiting
For the last melting of snow
The first butterflies.

Blue sky and sunshine
Clusters of bright ladybirds
The first bumbling bee.

Cold earth is turning
Slip sliding into the spring
As icicles melt.

Sunlight on hard frost
And deep within a green tree
Red admirals stir.

Late February
A calm day with warming sun
The first butterfly.

A single peacock
Out of winter's cold darkness
Basking in sunlight.

From its long sleeping
A comma with widespread wings
Soaking up the sun.

Between tree branches
A flicker of pale blue sky
As holly blues rest.

On a grey pathway
Wide wings of small tortoiseshells
Welcoming the sun.

Prefers hedge garlic
Warmed by rays of evening sun
Roosting orange tip.

From a coal blackness
To this large-eyed radiance
Peacock's open wings.

On such a dull day
Even a single small white
Brightens the landscape.

Along the leaf spine
Brimstone caterpillar rests
Green on green unseen.

Two commas spiral
Up and up into blue sky
Drifting with white clouds.

Resisting nectar
Painted ladies head inland
Millions of migrants.

Common blues in grass
A mundane name for these bright
Reflections of sky.

Sitting on a seat
I watch one caterpillar
On its own journey.

Two brown butterflies
Dancing beneath dappled leaves
Battling for sunspots.

White flags fluttering
Around the lavender beds
As small whites nectar.

The faintest tickle
As a comma leaves its leaf
And lands on my hand.

Skippers are skipping
From one grass stem to the next
That's what skippers do.

In the bramble glade
Large whites glide elegantly
Through shafts of sunlight.

Velvet smooth in the sun
The wings of summer peacocks
On their maiden flights.

An orange comma
Becomes a tattered dead leaf
Perfect camouflage.

Flying to the feast
Meadow browns land on the first
Sunlit buddleia.

Through the night's stillness
Ghost moths on hovering wings
Hunt the perfumed dark.

Loud group on the stage
And in a nearby green oak
Purple hairstreaks dance.

Three white butterflies
Dancing above fading leaves
And summer turning.

Juicy chunks of fruit
A dripping pile on the grass
Feeding butterflies.

On top of soft plum
The feeding red admiral
A deep crimson slash.

Two sunlit commas
On fermenting blackberries
Out of the wind's edge.

From a dark ivy
One red admiral seeks sun
And the last nectar.

Sheets of ice above
Deep blankets of settled snow
And butterflies sleep.

Index of Wildlife and Cultivated Plants

Illustrations are indicated by **bold** numbers

Alyssum 4, 48, 64
Aphids 26
Arabis 4, 48, 64
Ash 26
Aubretia 4, 48, 64
Badgers 73
Bee orchids 68
Bees 48, 73, 74
Bird's-foot trefoil 30, 66
Black medick 30
Black-tailed skimmer 59
Blackthorn 68
Bluebells 12, 61
Bramble 12, 14, 15, 16, **17**, 19, 20, 21, 26, 31, 32, 42, **47**, 48, 50, **54**, 56, 77, 78
Brimstone 16, **21**, 22, 51, 60, **61**, 62, 63, 68, **69**, 74, 76
Broad-bodied chaser **59**
Broom 4, **15**, 31, 41, 48, 52
Brown argus 28, **29**, 30, 60, 62, 63
Brown bent grass 70
Buddleia 14, 16, 26, 31, 32, 35, 42, 45, 48, 50, 54, 56, 64, 66, **67**, 68, 72, 78
Buckthorn 4, 16, 21, 22, 31, 48, 64, 66, 68, **69**
Buttercups 70
Cabbages 23, 24
Canadian pondweed 57
caeruleopunctata 28
Candy tuft 66
Celandines 12, 16
Chiffchaff 25
Cinnabar moth 12
Clouded yellow **51**, 60
Clover 12, 14, 16, 51
Cocksfoot grass 19
Comma 16, 17, **38**, 39, 60, 62, 63, 75, 76, 77, 78
Common blue 15, 28, **30**, 60, 62, 63, 66, 76
Common hemp nettle 14, 16
Corncockle 70
Corn marigold 70
Cotoneaster 64
Crab spider **57**
Creeping soft grass 19
Crested dog's-tail grass 70
Crocuses 12, 16
Crucifers 24
Daffodils 12
Dames' violet 66
Damselflies 73

Dandelions 12, 64
Dingy skipper 21, 60
Dock 27
Dogwood 68
Dragonflies 57, **59**, 73
Erigeron 14, 66
Essex skipper 18, **19**, 20, 60, 62, 63
Field maple 68
Fine bent grass 70
Foxes 73
Fungal infections 56
Garlic mustard 24, 64, 66
Gatekeeper 42, **44**, **45**, 47, 60, 62, 63, back cover
Geranium 29
Golden rod 66
Gorse 4, 15, 31, 48, 52
Grayling 42, **43**, 60, 62, 63, 73
Green hairstreak **52**, 60
Green-veined white 15, **23**, 24, 25, 60, 62, 63, 66
Guelder rose 68
Gulls 56
Harlequin ladybird 68
Hawkweed 12, 14, 15, 16, 45
Hawthorn 31, 52, 68
Hazel 68
Hedge brown - see gatekeeper
Hedge garlic 77
Hedgehogs 68
Hedge mustard 16, 24, 25
Hedges 12, 16, 17, 68, 70
Hibernation 16, 21, 32, 35, 36, 37, 38, 58, 66, 68, 74, 75, 78
Holly 14, 15, 16, 31, 66, 68
Holly blue 12, 14, 30, **31**, 32, 56, 60, 62, 63, 66, 75
Honesty 64
Honeydew 26
Honeysuckle 4, 48, 53, 66
Hop 38
Hornets 16, 33, 56
Hummingbird hawk moth 64, 72
Hyacinth 64
Ichneumon wasp 32
Insect 'hotels' 68
Ivy 12, 14, 16, 21, 31, 32, 56, 66, 68, 74, 78
Jack-by-the-hedge 14, 24
Knapweed 15
Ladybirds 68, 74
Lady's smock 14, 15, 24, 64

79

Large skipper 20, **58**, 60, 62, 63
Large white 21, **22**, 24, 60, 62, 63, 77
Lavender **9**, 12, 14, 24, 66, 76
Laurel 14, 15, 64
Leaf-cutter bees 73
Lizards 73
Lucerne 51
Marjoram 4, **48**, 64
Meadow brown 15, 42, 45, **46**, 47, 60, 62, 63, 78
Meadow foxtail grass 70
Meadow saffron 70
Mexican orange blossom 64
Michaelmas daisies 12, 66
Migration 24, 32, 34, 35, 36, 50, 51, 72, 76
Milk parsley 49
Mites 56
Moths 12, 67, 72, 78
Muntjac deer 73
Nasturtiums 23
Nettles 14, 16, 32, 35, 37, 38, 64
Oak 14, 16, 26, 27, 78
Orange tip front cover, 15, 24, **25**, 39, 60, 62, 63, 66, 76
Orwell, river 50
Painted lady 16, **34**, 51, 60, 62, 63, 72, 73, 76
Peacock frontispiece, 14, 16, 36, **37**, 39, 58, 60, 62, 63, 75, 76, 77
Phlox 14, 66
Predation 32, 33, 35, 37, 56, **57**, 58
Primroses 14
Primula 14
Privet 14
Purple hairstreak 14, 15, 16, **26**, 27, 60, 62, 63, 73, 78
Pyracantha 64
Quaking grass 70
Ragged robin 70
Ragwort 12, 14, 26
Red admiral 16, 17, 32, **33**, 35, 57, 60, 62, 63, **74**, 78
Red deadnettle 12, 16
Red fescue grass 70
Restharrow 30
Ringlet 14, 42, **47**, 60, 62, 63
Rockrose 29
Rough meadow grass 70
Rowan 52
Rue 49
Rye grass 70

Sap 33
Scabious 14
Sedum 66
Sheep's fescue grass 70
Silver-studded blue 60
Silver-washed fritillary 39, 54, **55**, 60
Silver-y 72
Skippers 77
Small copper 15, 27, **28**, 60, 62, 63
Small heath 60, 73
Small skipper 18, **19**, 20, 60, 62, 63, **77**
Small tortoiseshell 12, 16, 35, **36**, 56, 60, 62, 63, 66, **75**
Small white 22, **23**, 24, **57**, 60, 62, 63, 76
Smooth meadow grass 70
Snake's head fritillary 70
Snowdrops 12
Sorrel 27
Southern marsh orchid 70
Sparrowhawks 56
Speckled wood 14, 17, 39, **40**, 41, 60, 62, 63
Spiders 56, **57**
Spindle 68
Stag beetle clumps 16, 68
Storksbill 29
Stour, river 73
Sturmia bella fly 35
Summer jasmine 72
Sundew 24
Swallows 25, 73
Swallowtail 49, 50, 60
Sweet vernal grass 70
Sweet William 66
Sycamore 26
Thistles 34, 35
Thyme 4, **48**, 64
Timothy grass 19
Valerian 36, 66, **67**, 72
Verbena bonariensis 12, 14, 64, 72
Violets 54
Wall or **wall brown 41**, 42, 60, 62, 63
Wallflower Bowles' mauve 66, 72
Waterbirds 56
Watercress 24
Wavy hair-grass 70
White admiral 53, 54, 60, 66
White-letter hairstreak 27, 60, 73
Yarrow 12, 14, 15, 16, 27
Yellow rattle 70
Yorkshire fog grass 19